The Shape of
EARTH

Rebecca Woodbury, Ph.D., M.Ed.

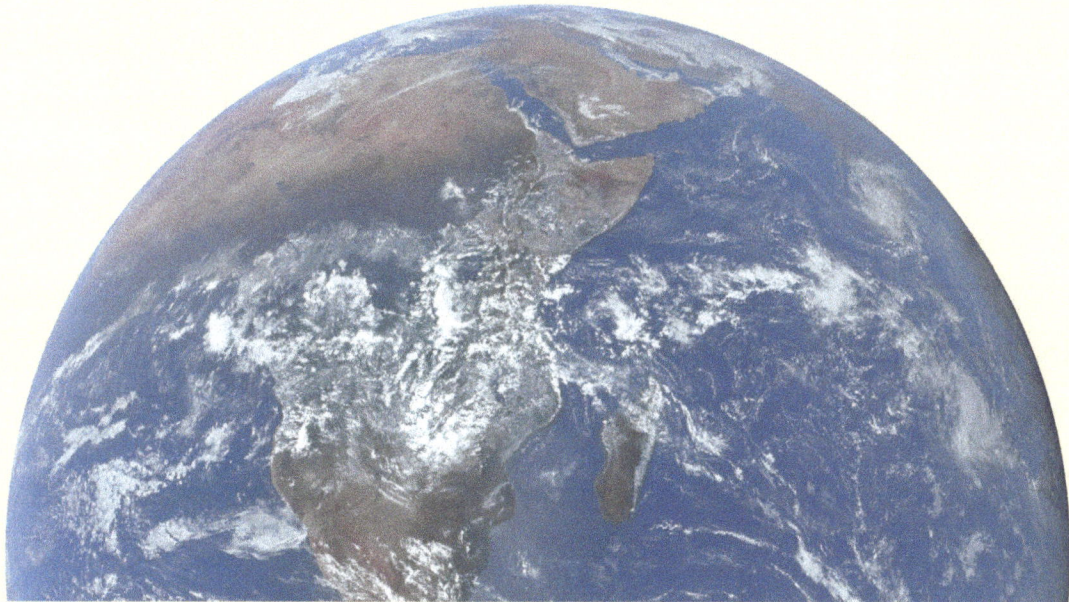

Gravitas Publications Inc.

The Shape of
EARTH

Illustrations: Janet Moneymaker

The Shape of Earth
ISBN 978-1-950415-32-8

Published by Gravitas Publications Inc.
Imprint: Real Science-4-Kids
www.gravitaspublications.com
www.realscience4kids.com

RS4K

Photo credits: Cover & Title Pg: NASA; P.11. Prikhodko, AsobeStock

What is the shape of Earth?

Hmm.
I wonder...

Is it round like a circle
and flat like a pancake?

Do you think the Earth is flat?

It seems like it.

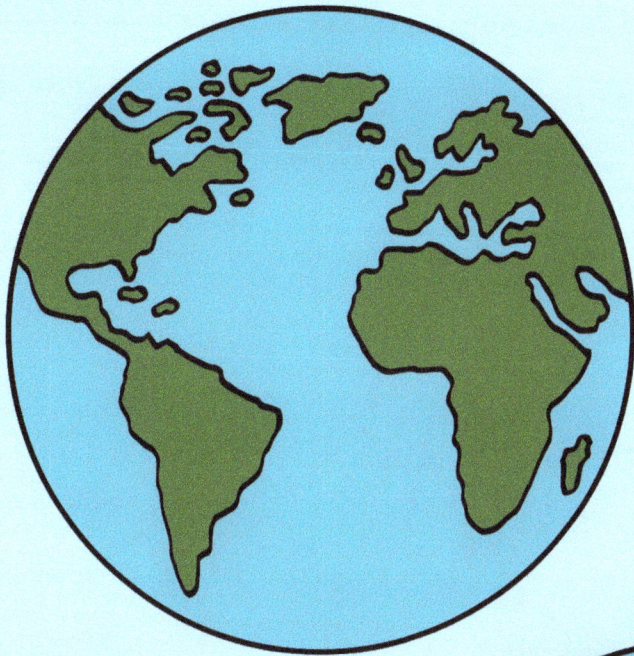

Is it square like a block?

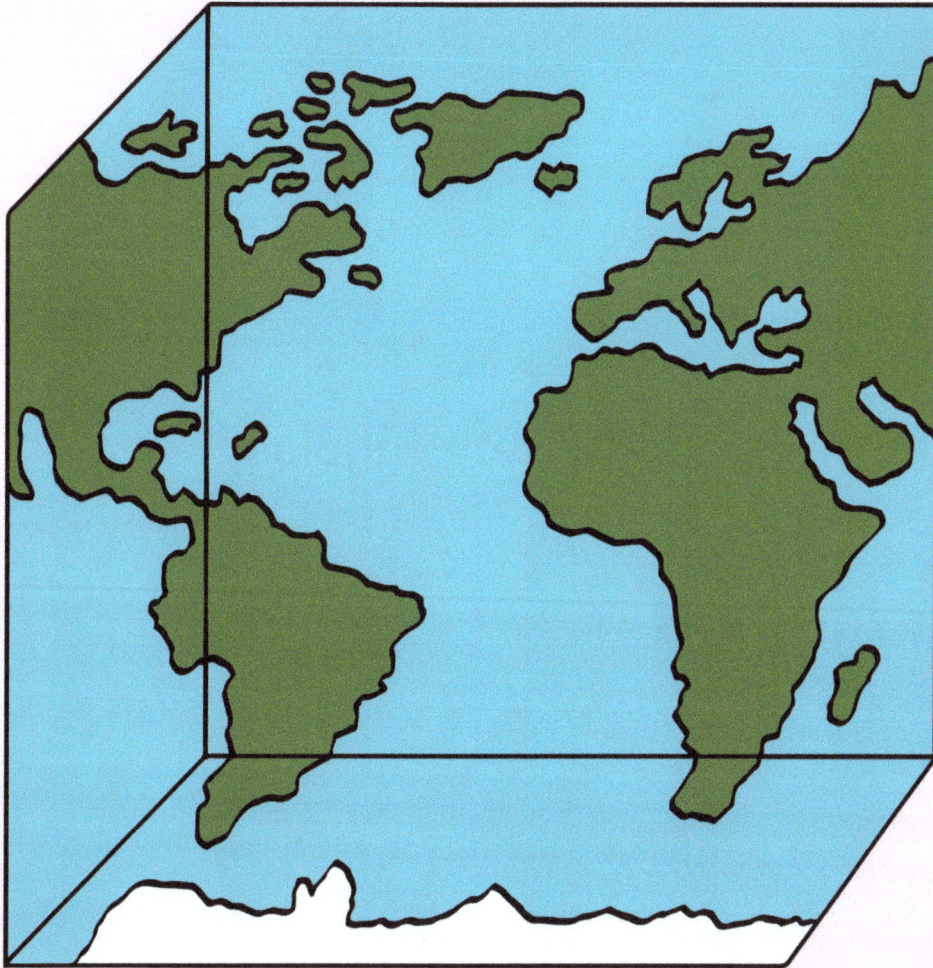

Is it shaped like a
pumpkin or an eggplant?

Do you think
Earth looks like a
piece of cheese?

Probably
not.

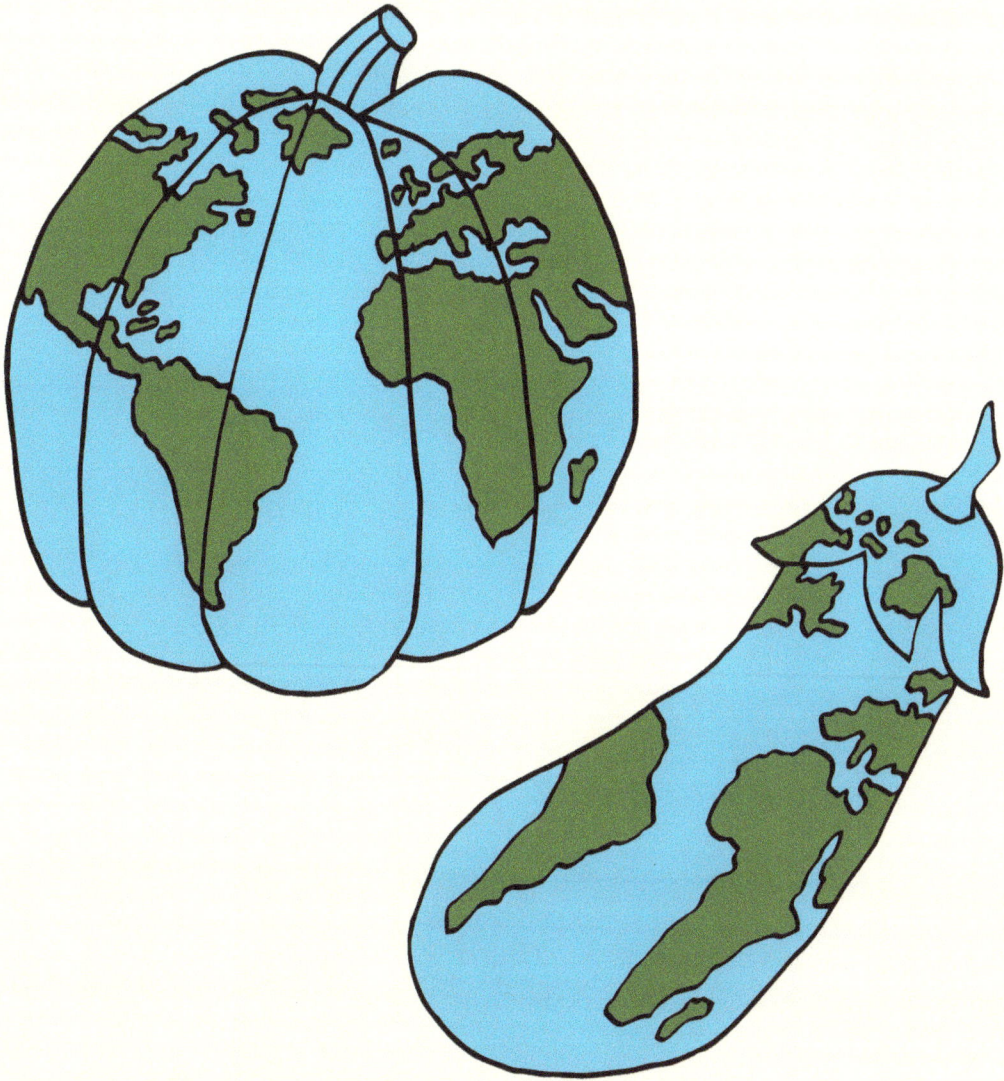

If you take a walk around your neighborhood, Earth seems flat.

It does seem flat.

If you climb up a hill and
go down the other side,
Earth still seems flat.

But! If you fly in a plane
or if you take a rocket to
the Moon, you can see
that Earth is not flat.

Earth is curved and shaped like a slightly flattened ball.

But why does Earth seem
flat when we walk on it?

Earth seems flat because it is so huge and we are so tiny. We only see a teeny bit of Earth's surface when we are not in an airplane or on a rocket. This makes Earth appear to be flat.

But we really live on a giant blue ball!

How to say science words

circle (SUR-kuhl)

curved (KURVD)

Earth (ERTH)

science (SIY-uhns)

shape (SHAYP)

www.ingramcontent.com/pod-product-compliance
Lightning Source LLC
Chambersburg PA
CBHW040153200326

41520CB00028B/7586